FLORA OF TROPICAL EAST AFRICA

ASPARAGACEAE

SEBSEBE DEMISSEW*

Perennial scandent or erect shrubs or subshrubs, branching; rhizomes sympodial from where the branches are growing; roots often swollen and fusiform. Spines present or absent; when present these are formed usually from the reduced leaves, occasionally from branches. Leaves normally reduced and scale-like, the assimilating function taken by modified green branches (cladodes); in some genera the branches are transformed into flattened, leaf-like cladodes (phylloclades). Cladodes solitary or fascicled, subulate, angled or linear. Inflorescence axillary or terminal, solitary, fascicled or assembled in racemes or "umbels". Flowers unisexual or bisexual, actinomorphic, small, erect or pendulous. Perianth with 6 tepals in two series, all similar in shape, free or fused at the base, white, cream, yellow or green. Stamens 6, in two series, fused to the perianth segments, present both in unisexual and bisexual flowers, non-functional in female unisexual flowers; filaments free from each other, anthers introrse, dorsifixed; pollen grains sulcate. Pistil with 3 carpels united to form a 2–3-locular ovary with 1–12 ovules in each locule; placentation axile; style short with capitate or lobed stigma. Fruit a globose berry with 1–2(–3) seeds. Seeds black, globose or truncate on one side, convex on the other.

Family represented by the genus *Asparagus* with two subgenera, subgenus *Asparagus* and subgenus *Myrsiphyllum* (S.T. Malcomber & Sebsebe Demissew in Kew Bull. 48(1): 63–78 (1993)). The members are widely distributed in the old world. Most species are found in arid tropical regions and Mediterranean climates.

ASPARAGUS

L., Sp. Pl. 1: 313 (1753); Jessop in Bothalia 9: 31–96 (1966); Kubutzki & Rudall in Kubutzki (ed.), Flowering Plants, Monocotyledons: Lilianae (except Orchidaceae) 3: 128 (1998)

Description as for the family.

Fewer than 300 species distributed throughout Africa, parts of Europe, Asia and Australia. Two subgenera and 23 native and two cultivated species are recognised in the Flora of Tropical East Africa.

Two cultivated species are:

A. officinalis L. (1753) is a cultivated herb to 2 m high, distinguished from the rest of the African native species by the flowers being unisexual. Recorded (in the Flora area) only from **T** 5: Dodoma District, C.D.A. Forest Nursery, 18 Oct. 1979, *Ruffo* 1220!; Dodoma, 29 Dec. 1979, *Ruffo* 1510! and reported in U.O.P.Z.: 133 (1949) from Zanzibar and Pemba.

* The National Herbarium, Science Faculty, Addis Ababa University, P. O. Box 3434, Addis Ababa, ETHIOPIA.

1

A. aethiopicus L. (1767) is native to South Africa. However, several cultivars of the species are grown in various parts of the world. The cultivar found in East Africa is cv. 'Sprengeri' (*A. sprengeri* Regel). P. Green (1986) has discussed the correct naming of *A. sprengeri*. It is common in pots and gardens and grows even in very hot areas if it is well watered. Plants are characterized by the drooping, loose and spreading branches. Obermeyer (1993) had kept this cultivar under *A. densiflorus* (Kunth) Jessop. Reported in U.O.P.Z.: 133 (1949) from Zanzibar and Pemba.

1. Cladodes subulate, filiform or linear, not leaf-like;
     flowers usually erect, rarely pendulous; filaments free
     (1. subgen. *Asparagus*) 2
   Cladodes flattened, leaf-like (phylloclade); flowers
     pendulous; filament connivent, forming a tube around
     the ovary (2. subgen. *Myrsiphyllum*) . . . .   23. *A. asparagoides*
2. Branches without spines 3
   Branches with spines 4
3. Cladodes in fascicles of 2–3, filiform; flowers pendulous;
     pedicel 4.5–8 mm long, articulated in lower half . . . . .   1. *A. virgatus*
   Cladodes in fascicles of 4–15; grooved; flowers erect;
     pedicel 2–3 mm long, articulated in the middle . . . . . .   5. *A. humilis*
4. Spines cauline in origin, placed in the axils of cladodes;
     branches terminating in spines . . . .   2. *A. suaveolens*
   Spines foliar in origin, placed below cladodes; branches
     not terminating in spines 5
5. Flowers solitary, paired or in clusters 6
   Flowers racemose or umbel-like in condensed inflorescences 19
6. Flowers solitary or paired 7
   Flowers 2–10 or more in fascicles 13
7. Cladodes turning black when dry; pedicel ± 2 mm long,
     articulated below perianth parts; spines 4–10 mm long   12. *A. schroederi*
   Cladodes remain green or pale green when dry; pedicels
     4–8 mm long, articulated in the middle or below;
     spines 1.5–3 mm long 8
8. Cladodes absent at anthesis 9
   Cladodes present at anthesis 10
9. Branches grooved or ridged; flowers axillary and terminal;
     spines on main stems and branches 1.5–2 mm long . . .   10. *A. denudatus*
   Branches terete; flowers axillary; spines on main stems
     and branches 3–5 mm long . . . .   11. *A. flagellaris*
10. Cladode fascicles close to each other giving the
     appearance of a 'bottle brush', up to 60 in a fascicle;
     pedicels articulated towards the base . . . .   3. *A. petersianus*
    Cladode fascicles spread apart, 3–25 in a fascicle; pedicel
     articulated in the middle or just below 11
11. Spines on main stems and branches 3–5 mm long;
     perianth white with pinkish tinge on the outside . . . .   11. *A. flagellaris*
    Spines on main stems and branches 1.5–2 mm long;
     perianth white 12
12. Young branches branching 4 ×; cladodes ± arranged in
     one plane; flowers terminal; pedicel articulated below
     the middle . . . .   8. *A. setaceus*
    Young branches only branching up to 3 ×; cladodes not
     as above; flowers axillary; pedicel articulated at the
     middle . . . .   9. *A. migeodii*
13. Stems and branches densely pubescent . . . .   6. *A. africanus*
    Stems and branches glabrous to puberulous 14

14. Cladodes absent for most of the time, when present in
     fascicles of 2–3 . . . .   10. *A. denudatus*
     Cladodes present all the time (?persistent) in fascicles of ≥ 4 15
15. Branches zigzagging; young stems white, ribbed; cladode
     fascicles straight and fan-shaped . . . .   4. *A. laricinus*
     Branches straight, not zigzagging; young stems greenish,
     terete or striated not ribbed; cladode fascicles commonly
     curved, not fan-shaped 16
16. Young branches branching 4 ×; cladodes fine, ± arranged
     in one plane; flowers terminal . . . .   8. *A. setaceus*
     Young branches only branching up to 3 ×; cladodes not as
     above; flowers axillary and/or terminal 17
17. Older stems not peeling off, terminal branches glabrous
     to puberulous, commonly without spines; cladodes
     3–15 mm long, rounded or angled 18
     Older stems peeling off; terminal branches glabrous and
     always with spines; cladodes 15–30 mm long, flattened
     or with grooves above . . . .   7. *A. scaberulus*
18. Cladodes on branchlets 'feather-like', pale green; spines
     on main stems ± 2 mm long; flowers axillary, solitary or
     in fascicles of 2–3; pedicel articulated in the middle . . .   9. *A. migeodii*
     Cladodes on branchlets not as above, green, spines on
     main stems 3–5 mm long; flowers axillary and terminal,
     in fascicles of 2–10(–35); pedicels articulated below the
     middle . . . .   6. *A. africanus*
19. Cladodes subulate or only slightly flattened, 8–35 × <1 mm;
     bracts ovate, 2–4 mm long 20
     Cladodes linear, flattened 7–85 × 1–5 mm; bracts lanceolate,
     1–1.5 mm long 24
20. Branches peeling off; racemes condensed; flowers easily
     fall off (caducous) from the pedicels . . . .   14. *A. leptocladodius*
     Branches not as above; raceme simple, or branching, not
     condensed; flowers persistent on the pedicels 21
21. Racemes 2–6; bracts 1.5–4 mm long; pedicel articulated
     in the middle or below . . . .   13. *A. racemosus*
     Racemes 1(–2); bracts 1–2 mm long; pedicel articulated
     above the middle or just below the perianth 22
22. Young branches grey, scabrid to puberulous; anthers black 23
     Young branches pale brown, glabrous, smooth; anthers
     cream to yellowish . . . .   17. *A. buchananii*
23. Cladodes <0.5 mm wide; inflorescence axis thin, not
     winged; perianth elliptic, ± 3 × 1 mm; pedicels 1.5–
     2.5 mm long . . . .   15. *A. aspergillus*
     Cladodes 0.5–1 mm wide; inflorescence axis thickened,
     slightly winged; perianth broadly elliptic, 2.5 × 1.5 mm;
     pedicels 0.5–1 mm long . . . .   16. *A. rogersii*
24. Cladodes (25–)30–105 mm long, 2.5–5 mm wide, mid-vein
     distinct . . . .   18. *A. falcatus*
     Cladodes 10–25 mm long, 0.5–2 mm wide, mid-vein
     indistinct 25
25. Inflorescence modified branchlets; outer perianth
     segments ciliate; fruit 9–10 mm in diameter . . . .   19. *A. natalensis*
     Inflorescence simple raceme; outer perianth segments
     entire, not ciliate; fruit 5–7 mm in diameter 26

26. Inflorescences 4–10 cm long; pedicels articulated near
the middle or below . . . .   22. *A. usambarensis*
Inflorescences 1.5–2.5 cm long; pedicels articulated
above the middle 27
27. Branches terete, not striated; grey; pedicels 3–4 mm long .   20. *A. aridicola*
Branches angled and striated; pedicels 1–1.5 mm long . . .   21. *A. faulkneri*

Subgenus ASPARAGUS

Asparagopsis Kunth in Abh. Akad. Berl. 35 (1842); Enum. Pl. 5: 76 (1850)
Protasparagus Oberm. in S. Afr. J. Bot. 2: 243–244 (1983)

1. **Asparagus virgatus** *Baker* in Saunders Ref. Bot. 3: t. 214 (1870); Baker in J.L.S. 14: 606 (1875) & in Fl. Cap. 6: 259 (1896); Jessop in Bothalia 9: 53 (1966). Type: illustration in Ref. Bot. 3: t. 214 (1870), iconotype

Stiff, erect, shrubs 30 cm to 1 m high; stems solitary or 2–4, thin, branching in ± regular fashion in the upper half or so, angled, ± without spines, sometimes without cladodes, glabrous. Cladodes in fascicles of 2–3, filiform, angled, 4–12 mm long, unequal. Flowers solitary, pendulous, along the branches; bracts ovate, 1 × 0.5 mm, acute at apex; pedicel 4.5–8 mm long, articulated in lower half, elongating to 10 mm long in fruit. Tepals white to yellowish, slightly unequal, 3–4.5 × 0.5–1 mm, outer ones smaller than the inner ones; stamens 6, attached to the perianth; ovary globose, 3-locular with 4(–6) ovules in each locule; style exserted at anthesis, 2–3 mm long, as long as the ovary. Berry greenish when young, turning yellow to red at maturity, 1–2-lobed, 4–7 mm in diameter; seeds black.

Tanzania. Iringa District: Mt Msalaba E, 21 Mar. 1991, *Gereau & Kayombo* 4445!; Iringa District: Irundi Hill, N of Lugoda, 18 May 1990, *Carter, Abdallah & Newton* 2299!; Songea District: Matengo Hills, 23 May 1956, *Milne-Redhead & Taylor* 10420!
Distr. **T** 7, 8; Angola, Zambia, Malawi, Mozambique, Zimbabwe, Namibia and South Africa
Hab. Secondary bushland and rocky slopes with scattered *Protea* etc.; 1600–2150 m

Syn. *Asparagus virgatus* Baker var. *capillaris* Baker in J.L.S. 14: 606 (1875); Baker in Fl. Cap. 6: 259 (1896). Type: South Africa, Cape Province, Caffraria, *Cooper* 202 (K!, iso.)
    Protasparagus virgatus (Baker) Oberm., F.S.A. 5(3): 31 (1992)

Note. The characteristic features of the species are the non-spiny and nodding nature of the branches, and solitary flowers.

2. **Asparagus suaveolens** *Burch.*, Travels Int. S. Africa 2: 226 (1824); Pole Evans in Fl. Pl. S. Afr. 11: t. 409 (1931); Jessop in Bothalia 9: 45 (1966). Type: South Africa, Cape Province, Griquatown, *Burchell* 1956 (K!, holo.)

Erect shrub, up to 1 m high; stems erect, straight or slightly zigzagging, with short internodes, glabrous; branches and main stems with spines 6–9 mm long. Cladodes solitary or in fascicles of 2–6, subulate, 3–6(–19) mm long. Flowers solitary or in fascicles of 2–3 on terminal branches; bracts overlapping, broadly elliptic, 1 × 0.5 mm, rounded to acute at apex; pedicel 1–5(–10) mm long, articulated near base. Tepals narrowly obovate, 1.5–3.5 mm long, white with a dark midrib; ovary obovoid, 3-locular with 4–6 ovules in each locule; style and stigmas short. Berry black, ± 5 mm in diameter, the dry perianth persistent, 1–3-seeded.

Kenya. Kericho District: Sotik, 27 Jan. 1965, *Parker* EA 13143!; Forest near Nairobi, Recd. Aug. 1912, *Mason* s.n.!
Tanzania. Musoma District: Serengeti, N of Bolgonja, 12 Apr. 1972, *Schmidt* 417!; Masai District: 9 km from Lokisale on Arusha Road, 11 June 1965, *Leippert* 5871!

Distr. **K** 4, 5; **T** 1, 2; widespread and common all over southern Africa except northern Namibia and northern Botswana

Hab. Few data; termite mound in *Acacia* wooded grassland, near forest and along roads; 1500–2000 m

Syn. *Asparagus triacanthus* Roem. & Schult., Syst. Veg. ed. 7, 7(1): 334 (1829). Type: South Africa, Cape, *Lichtenstein* in Herb. Willd. no. 6693 (B-W!, holo.)

Asparagopsis triacanthus (Roem. & Schult.) Kunth, Enum. Pl. 5: 91 (1850)

A. zeyheri Kunth, Enum. Pl. 5: 92 (1850). Type: South Africa, without locality, *Zeyher* s.n. (B!, holo.)

Asparagus spinosissimus Kuntze, Rev. Gen. Pl. 3(2): 315 (1898). Type: South Africa, Cape Province, Cathcart, *Kuntze* s.n. (NY!, holo.)

A. omahekensis Krause in E.J. 51: 447 (1914). Type: Namibia, Omaheke between Gobabis and Oas, *Dinter* 2711 (B!, holo.)

A. intangibilis Dinter in F.R. 29: 269 (1931). Type: Namibia, 'Grosse Karasberge', *Dinter* 5168 (B!, holo.)

Protasparagus suaveolens (Burch.) Oberm. in S. Afr. Journ. Bot. 2: 244 (1983)

3. **Asparagus petersianus** *Kunth*, Enum. Pl. 5: 72 (1850); Garcke in Peters, Reise Mossamb. Bot.: 520 (1864); Baker in J.L.S. 14: 614 (1875). Type: Mozambique, mouths of Zambesi R., *Peters* s.n. (B!, holo.)

Scandent or trailing shrub, glabrous. Branches many, terete, smooth. Spines at the base of the branches, 2–3 mm long, curved downwards; those at the base of the cladodes smaller, ± 1 mm long. Cladode fascicles close to each other on terminal or side branches, giving the appearance of a bottle brush; cladodes in fascicles of 10–60 or more, filiform, slightly curved, unequal, 7–10 mm long. Flowers solitary; bracts scarious, ± 0.5 mm long; pedicel ± 5 mm long in fruit, articulated close to the base. Tepals not seen. Berry orange, 7–10 mm in diameter.

Tanzania. Masai District: Mtama–Masasi road, 20 km from Masasi, 17 Mar. 1963, *Richards* 17892!; Uzaramo District: 34 km S of Dar es Salaam on road to Kilwa, 17 Feb. 1970, *Harris & Tadros* 4169!

Distr. **T** 2, 6; Mozambique

Hab. Secondary scrub woodland and relict forest; 0–600 m

4. **Asparagus laricinus** *Burch.*, Travels Int. S. Africa 1: 537 (1822); Bresler, Dissert. inaug.: 40 (1826); Roem. & Schult., Syst. Veg. 7(1): 337 (1829); Kunth, Enum. Pl. 5: 75 (1850); Baker in J.L.S. 14: 620 (1875); Baker in Fl. Cap. 6: 267 (1896); Sölch et al., Fl. Südwest-Afr.: 31 (1961); Jessop in Bothalia 9: 60 (1966). Type: South Africa, Cape Province, Hay between Griquatown and Wittewater, *Burchell* 1871 (K!, holo.)

Spiny shrubs 1–3 m high, erect or climbing or sometimes trailing; young stems and branches whitish, ribbed, minutely hispidulous or glabrous, turning brown and smooth with age; spines short, hard, straight or slightly curved, on stems and below branches 3.5–6(–8) mm, below cladode fascicles 1.5–2 mm long. Cladodes up to 10 in a fascicle, filiform, 8–15(–20) mm long when mature, fairly equal in length. Flowers 1–8, on outside of cladode fascicles; bracts overlapping, ± membranous, ovate, 1.5 × 1 mm, rounded at apex; pedicel 5–6 mm long, articulated in the lower half, below the middle. Tepals 2.5–4 mm long, white; stamens with red or orange anthers; ovary 3-locular with 5–6 ovules in each locule; style short, ± 1 mm long (incl. stigma) with 3 short, spreading stigmas. Berry red, 6–8 mm in diameter, 1-seeded.

Tanzania. Sumbawanga District: N Sanga Forest, 31 Nov. 1961, *Robinson* 4838!; Mbeya District: Poroto Mts, 16 May 1957, *Richards* 9756!; Iringa District: E slope above Lake Ngwazi, 1 km above Ngwazi Dam, 11 Aug. 1971, *Perdue & Kibuwa* 11025!

Distr. **T** 4, 7; Angola, Zambia, Zimbabwe, Botswana, Namibia and South Africa

Hab. Grassland, flooded grassland, bushland and thicket, forest margins; (750–)1700–2400 m

Syn. *Asparagus angolensis* Baker in Trans. Linn. Soc. ser. 2, 1: 254 (1878). Type: Angola, Huilla,
    Welwitsch 3879 (K!, holo.)
    Protasparagus laricinus (Burch.) Oberm. in S. Afr. Journ. Bot. 2: 244 (1983)

Note. *Bidgood et al.* 3550 from **T** 4 is included here due to its ribbed branches and red-colored
    fruits, but differs by having wider cladodes ± 0.5 mm and the absence of spines below
    cladode fascicles. More material from the area is needed to establish the correct identity of
    the specimen.

    5. **Asparagus humilis** *Engler* in E.J. 45: 154 (1910). Type: Tanzania, Uzaramo
District, 'Light House Island' near Dar es Salaam, *Engler* 2110 (B!, holo.)

    Perennial herb or subshrub, erect or prostrate, 10–50 cm high, sometimes
attaining 1.8 m high when supported by bushes; branches grooved, glabrous,
sometimes papillate, without spines. Cladodes in fascicles of 3–15, grooved, unequal,
4–15 mm long, spine-like at the tip. Flowers solitary or paired in cladode fascicle on
terminal branches; bracts elliptic, 1 × 0.5 mm, acute at apex; pedicel 2–3 mm long,
articulated ± in the middle. Tepals white, 3–3.5 mm long; ovary 3-locular; style ±
0.7 mm long with 3-branched stigma. Berry red, 5–7 mm in diameter.

Kenya. Kilifi District: Arabuko-Sokoke, 1 Jan. 1992, *Luke* 3029!; Kilifi area, 6 June 1981, *de
    Meester* 19/79!
Tanzania. Uzaramo District: Dar es Salaam, May 1952, *Revell* 119! & Bongoyo Island, 2 July
    1969, *Mwasumbi* 10550!; Mikindani District: Mtwara–Mikindani Road, 11 Mar. 1963, *Richards*
    17840!; Zanzibar, Kizimkazi, 11 Jan. 1931, *Vaughan* 1823!
Distr. **K** 7; **T** 6, 8; **Z**; Mozambique
Hab. Coral outcrops and salt flats near mangrove; 0–15 m

    6. **Asparagus africanus** *Lam.*, Encycl. Meth. Bot. 1: 295 (1783); Bresler, Dissert.
inaug.: 9 (1826); Roem. & Schult., Syst. Veg. 7: 331 (1829); Baker in J.L.S. 14: 619
(1875) & Fl. Cap. 6: 265 (1896); Marloth, Fl. S. Afr. 4: pl. 20 (1915); Salter in Fl. Cape
Penins.: 175 (1950); Kies in Bothalia 6: 177 (1951); F.P.U.: 203, fig. 132 (1962);
F.W.T.A. 3: 94, fig. 349 (1968); Blundell, Wild Fl. E. Afr.: fig. 245 (1987); U.K.W.F.:
311 (1994); Sebsebe Demissew in Fl. Somalia 4: 25 (1995) & Fl. Eth. & Erit. 6: 68
(1997). Type: South Africa, Cape without precise locality, *Sonnerat* s.n. (P!, holo.)

    Erect or climbing or scrambling shrub to 4 m; branches terete to angled, glabrous
to pubescent, with spines 3–5 mm long; terminal branches with or without spines.
Cladodes fasciculate, 5–25, subulate, stiff or flexible, 3–15 mm long. Flowers in
fascicles of 2–10(–35), axillary and terminal; bracts lanceolate, ± 1.5 mm long, falling
off quickly; pedicels 3–8(–10) mm long, articulated below the middle. Tepals white
to cream, ± equal, 3–4 mm long, entire; stamens shorter than the perianth, anthers
yellow; ovary 3-locular with 4–8 ovules in each locule; style 1 mm long, 3-branched.
Berry red, 5–6 mm in diameter, 1-seeded. Fig. 1: 1–3 (page 7).

Stems and branches glabrous to puberulous; cladodes present
    at anthesis . . . .   var. **africanus**
Stems and branches densely pubescent; cladodes usually absent
    at anthesis . . . .   var. **puberulus**

    var. **africanus**

Uganda. Kigezi District: Mushongero, 21 Aug. 1938, *Thomas* 2392!; Mbale District: Elgon, Jan.
    1918, *Dummer* 3737!
Kenya. Narok District: OlTarakwai, 18 Aug. 1961, *Glover et al.* 2483!; Naivasha District: Lake
    Naivasha, 15 Oct. 1964, *E.F. Polhill* 90A!; Kiambu District: Dagoretti corner along the
    Nairobi–Nakuru road, 2 Jan. 1972, *Safiri, Kibui & Njunge* 3!

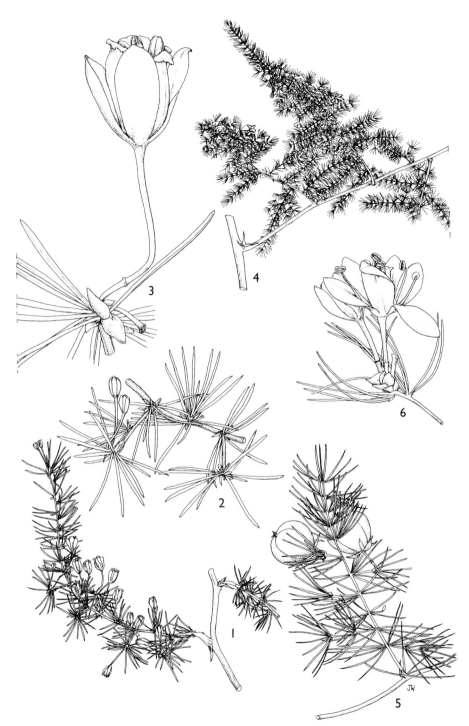

FIG. 1. *ASPARAGUS AFRICANUS* — **1**, flowering branch, × ²/₃; **2**, branch with cladodes, × 2; **3**, pedicel with flower, × 5. *ASPARAGUS SETACEUS* — **4**, flowering branch, × ²/₃; **5**, branch with cladodes and fruit, × 2; **6**, pedicel with flower, × 5. 1–3 from *Hedberg* 1044. 4, 6 from *Richards & Arasululu* 28998; 5 from *Bridson* 574. Drawn by Juliet Williamson.

TANZANIA. Arusha District: Ngare Nanyuki, 25 Feb 1968, *Richards* 23110!; Singida District: Minyighe Forest, 13 Dec. 2001, *Gereau & Mawi* 6702!; Iringa District: Mafinga [Sao Hill], 24 Nov. 1980, *Ruffo* 1591!; Zanzibar, Unguja Ukuu, 24 Aug. 1962, *Faulkner* 3092!
DISTR. **U** 2, 3; **K** 1–4, 6, 7; **T** 1–8; **Z**; Sudan, Ethiopia, Eritrea S to South Africa; also Arabia to India
HAB. Bushland, thicket, forest margins and grassland, secondary and ruderal vegetation; 0–3000(–3500) m

SYN. *Asparagus mitis* A.Rich., Tent. Fl. Abyss. 2: 319 (1851). Type: Ethiopia, Tigray, Tcheleukote, *Petit* s.n. (P, syn.); Adwa, *Schimper* 296 (BR!, K!, syn.)
     A. cooperi Baker in Gard. Chron. 1: 818 (1874). Type: South Africa, Cape Province, without precise locality, *Cooper* s.n. ex Hort. Saunders 7/71 *Asparagus* 1449 (K!, syn.); Cape Province, Boschberg, *MacOwan* 1810 (K!, syn.)
     A. asiaticus sensu Baker in J.L.S. 14: 618 (1875) & Fl. Cap. 6: 265 (1896) & Cuf., E.P.A. 41 (3): 1562 (1971), *non* L. (1753)
     A. irregularis Baker in J.L.S. 14: 620 (1875). Type: Malawi, foot of Chiradzulu, 3 Oct. 1859, *Kirk* s.n. (K!, holo.)
     A. judtii Schinz in Bull. Herb. Boiss. 1, 4, app. III: 44 (1896); Sölch, Beitr. Fl. Südwest-Afr. 37 (1961). Type: Namibia, Hoachanas, *Fleck* 901 (Z!, holo.)
     A. conglomeratus Baker in F.T.A. 7: 428 (1898). Type: Botswana, Ngamiland, Kwebe, *Lugard* 52 (K!, holo.)
     A. asiaticus L. var. *ellenbeckianus* Engler in Sitz. K. Preuss. Akad. Berlin 40: 737 (1906), *nom. nud.* Type: Ethiopia, Bale, Ladjo, *Ellenbeck* s.n. (B†, holo.)
     A. asiaticus L. var. *mitis* (A.Rich.) Chiov. in Nuovo Giorn. Bot. Ital. 26: 166 (1919)
     A. sidamensis Cufodontis in Senck. Biol. 1: 245 (1969). Type: Ethiopia, Sidamo, Mt Damot near Soddo, *Kuls* 453 (FR!, holo.)
     Protasparagus africanus (Lam.) Oberm. in S. Afr. Journ. Bot. 2: 243 (1983)
     P. cooperi (Baker) Oberm. in S. Afr. Journ. Bot. 2: 243 (1983)

NOTE. Most of the synonyms from South Africa, Botswana and Namibia indicated here and in Jessop (1966) under *A. africanus* are kept under *P. cooperi* by Obermeyer in F.S.A. 52(3): 33–34 (1992). *A. cooperi* sensu Obermeyer represents the forms with shorter spines while *A. africanus* is restricted to forms with sharp brown spines 7–10 mm long. However, the characters given are variable and hence only one variable species, *A. africanus* is recognized here.

     var. **puberulus** (*Baker*) *Sebsebe* **stat. & comb. nov**. Type: Malawi, Mangomoro, Manganja Hills, Oct. 1861, *Meller* s.n. (K!, holo.)

KENYA. Machakos/Kitui Districts: "Ukambani", 1893–94, *Scott Elliot* 6490!
TANZANIA. Shinyanga District: Shinyanga, Nov. 1938, *Koritschoner* 1826!; Manyoni District: Chaya, 20 Nov. 1961, *Semsei* 3427!; Masasi District: 3 km N of Nangomba village on Masasi–Songea road, 10 Nov. 1978, *Magogo & Innes* 484!
DISTR. **K** 4; **T** 1, 2, 4–8; also in Angola, Zambia, Mozambique, Zimbabwe and Botswana
HAB. Grassland, *Brachystegia* woodland, *Combretum* bushland, common on termite mounds; 100–2400 m

SYN. *A. puberulus* Baker in J.L.S. 14: 618 (1875)
     A. pubescens Baker in Trans. Linn. Soc., ser 2, 1: 254 (1878); F.W.T.A. 3: 93 (1968). Type: Angola, Huila District, near Catumba, *Welwitsch* 3878 (K!, holo., BM!, iso.)
     A. pilosus Baker in J.L.S. 14: 610 (1875). Type: Botswana, near Lake Ngami, *McCabe* 15 (K!, holo.)
     A. shirensis Baker in F.T.A. 7: 427 (1898). Type: ?Malawi, 1891, *Buchanan* 1003 (K!, holo.)

     7. **Asparagus scaberulus** *A.Rich.*, Tent. Fl. Abyss. 2: 320 (1851); Sebsebe Demissew in Fl. Somalia 4: 25 (1995) & Fl. Eth. & Erit. 6: 68 (1997). Type: Ethiopia, Tigray, Choho, *Quartin-Dillon* s.n. (P!, holo. & iso.)

     Erect to climbing shrub to 2 m high; branches purplish brown greyish, terete, smooth or lined, sometimes peeling, glabrous; spines 1–3 mm long, curved downwards, also on the terminal branches. Cladodes in fascicles of 4–25(–35), flexible, straight or bent, 15–30 mm long, flattened, angled, sometimes forming grooves on the upper side. Flowers in fascicles of (2–)3–6, axillary or terminal; bracts ovate, ± 1 mm long, membranous; pedicel 4–10 mm long, articulated at the middle or below. Tepals white, ± equal, 3–4 mm long; stamens shorter than the perianth, anthers yellow; ovary 3-locular with 5 ovules in each locule; style ± 1 mm long with 3-branched stigma. Berry red, 4–5 mm in diameter, 1-seeded.

UGANDA. Karamoja District: near Moroto, 24 June 1962, *Miller* 582!
KENYA. Northern Frontier Province: Dandu, 3 Apr. 1952, *Gillett* 12682!; Turkana District: Lorengipe, 9 Apr. 1954, *Hemming* 260!; Tana River District: Galole, Nov. 1964, *Makin* in EA 13055!
TANZANIA. Masai District: Ngorongoro Conservation area, Olduwai Camp, 21 Oct. 1989, *Chuwa* 2877!; Masai District: 13 km S of Namanga, 12 Dec. 1959, *Verdcourt* 2523!; Kondoa District: near Mangaloma, 21 Dec. 1927, *Burtt* 868!
DISTR. U 1; K 1, 2, 1/4, 6, 7, T 1, 2, 5; Ethiopia, Eritrea and Somalia
HAB. *Acacia–Commiphora* woodland, open *Acacia* woodland, grassland, bushed grassland, thicket; (60–)350–1200(–1700) m

SYN. *A. asiaticus* L. var. *scaberulus* (Rich.) Engler, Hochgebirgsfl. trop. Afr.: 169 (1892)
   A. exuvialis sensu Agnew, U.K.W.F. ed. 2: 311 (1994), *non* Burch.

NOTE. *Newbould* 2850, collected from K 1, is included under this species, but stems are pubescent – unusual for this species.

8. **Asparagus setaceus** (*Kunth*) *Jessop* in Bothalia 9: 51 (1966); U.K.W.F.: 311 (1994); Sebsebe Demissew in Fl. Eth. & Erit. 6: 68 (1997). Type: South Africa, *Drège* 8584 (B†, holo.; KIEL!, lecto.)

Climbing shrub to 8 m high; branches branching to 4 orders, terete or grooved, with spines 2–3 mm long mainly on the main branches, glabrous. Cladodes arranged in ± one plane, in fascicles of 4–25, linear, fine, 3–6(–10) mm long. Flowers solitary or fasciculate, 2–3 together on terminal branches; bracts minute, falling more or less quickly; pedicels 4–8 mm long, articulated at the middle or below. Tepals white, ± equal, ± 3 mm long; stamens shorter than the perianth; anthers yellow; ovary 3-locular with 6–8 ovules in each locule; style ± 1 mm long, 3-branched. Berry red, 6–8(–10) mm in diameter, 1–3-seeded. Seeds black. Fig. 1: 4–6 (page 7).

KENYA. Meru District: upper Imenti Forest, 0°5'N 37°37'E, 28–29 June 1974, *R.B & A.J. Faden* 74/903!; Masai District: Kajiado, Ol Doinyo Orok, 27 Apr. 2000, *Massawe et al.* 613!; Teita District: Taita Hills, East of Mwandingo Forest, 6 Nov. 1998, *Mwachala et al.* EW820!
TANZANIA. Lushoto District: Chambogo Forest Reserve, 19 July 1987, *Kisena* 441! Ufipa District: Mbisi [Mbizi] Forest Reserve, 29 Oct. 1987, *Ruffo & Kisena* 2826!; Mbeya District: near Tewe village in Undali [Bundalli] Hills, 6 Nov. 1966, *Gillett* 17580!; Zanzibar: Ufufuma, 7 Feb. 1929, *Greenway* 1369!
DISTR. K 3, 4, 6, 7; T 2–7; Z; Ethiopia, Eritrea, Zambia, Malawi, Zimbabwe, South Africa
HAB. Forest and forest margins, rarely in thicket or dense bushland on coral; 0–350 m near coast and 750–2300 m inland

SYN. *Asparagopsis setacea* Kunth, Enum. Pl. 5: 82 (1850)
   Asparagus plumosus Baker in J.L.S. 14: 613 (1875); Planchon in Flor. des Serres ser. 2, 23: t. 2413–2414 (1880); Baker in Fl. Cap. 6: 260 (1896); U.O.P.Z.: 133 (1949). Syntypes: South Africa, Port Natal, *Drège* 4482 (K!, syn.); "Kaffraria", *Cooper* 202 (K!, syn.); Natal, *Gerrard & McKen* 754 (TCD!, syn.)
   A. zanzibaricus Baker in J.L.S. 14: 614 (1875). Type: Tanzania, Zanzibar Island, *Hildebrand* 1048 (BM!, holo.; K!, iso.)
   A. asiaticus L. var. *amharicus* Pic.-Serm. in Miss. Stud. Lago Tana, Ric. Bot. 1, 7: 194 (1951). Type: Ethiopia, Gojam, Zeghe Peninsula, *Pichi-Sermolli* 2027 (FT!, holo.; K!, iso.)
   Protasparagus plumosus (Baker) Oberm. in S. Afr. Journ. Bot. 2: 244 (1983) & F.S.A. 5(3): 59 (1992)
   P. setaceus (Kunth) Oberm. in S. Afr. Journ. Bot. 2: 244 (1983) & F.S.A. 5(3): 58 (1992)

9. **Asparagus migeodii** *Sebsebe* **sp. nov** *A. setacei* (Kunth) Jessop surculis terminalibus plumosis similis sed habitus erecto non scandenti, floribus axillaribus non terminalis differt. Type: Tanzania, Lindi District, Tendaguru, *Migeod* 468 (BM!, holo.)

Erect perennial herb; stem smooth, glabrous; spines on main stems ± 2 mm long. Cladodes in fascicles of 5–10 with white stipules at the base, linear, fine, 5–6 mm long. Flowers solitary turning brown within the cladode fascicles; outer stipules similar to bracts, ovate, 1 × 0.5 mm, acute at apex, inner ones lanceolate with acuminate apex; bracts ovate, 2 × 1.5 mm, acute at apex; pedicels 4–5 mm long, articulated ± at the middle. Tepals white, ± equal, 2–3 mm long; stamens shorter than the perianth; anthers yellow; ovary 3-locular, with 6–8 ovules in each locule; style ± 1 mm long, 3-branched. Berry red, 5–6 mm in diameter, 1–2-seeded.

TANZANIA. Lindi District: Tendaguru, 7 Feb. 1926, *Migeod* 82A! & 14 May 1929, *Migeod* 468!
HAB. Wooded grassland; 210 m
DISTR. **T** 8; Zambia

NOTE. The new species is superficially similar to *A. setaceus*, in having fine and feathery-like cladodes, but differs by the ± erect habit and axillary flowers. In contrast, *A. setaceus* has a climbing habit and terminal flowers. It is also related to *A. africanus*, but it is distinguished by the cladodes on branchlets 'feather-like' and pale green; spines on main stems ± 2 mm long; flowers axillary, and solitary or paired. In contrast, *A. africanus* has cladodes on branchlets not as above, green, spines on main stems 3–5 mm long; flowers in fascicles of 5–25, axillary and/or terminal.

10. **Asparagus denudatus** (*Kunth*) *Baker* in J.L.S. 606 (1875). Type: South Africa, Cape Province, Queens town, Stormberg, *Drège* 3533 (B!, holo.; BOL!, lecto.; K!, isolecto.)

Erect branching shrub, 1–1.5 m high, leafless for most of the time; branches grooved or ridged, with spines 1.5–2 mm long, glabrous. Cladodes in fascicles of 2–3, 6–15 mm long (2–4 mm long in South Africa). Flowers solitary or in 2–5 fascicles, axillary and terminal; bracts ovate, 2 × 1 mm, obtuse at apex; pedicels 4–7 mm long, articulated in the middle or below. Tepals white, 2.5–3.5 × 1–1.5 mm; ovary 2-locular with 3–4 ovules in each locule. Berry red or black, 4–5 mm in diameter, 1-seeded.

SYN. *Asparagopsis denudatus* Kunth, Enum. Pl. 5: 82 (1850), as *denudata*
    Protasparagus denudatus (Kunth) Oberm. in S. Afr. Journ. Bot. 2: 243 (1983)

subsp. **nudicaulis** (*Baker*) *Sebsebe* **comb. & stat. nov**. Type: Tanzania, Moshi District: in the steppes near Lake Chala [Dschalla], *Volkens* 1806 (K!, holo.)

Branchlets almost 45° to the main stem; branching internodes 30–60 cm long; cladodes 6–15 mm long; pedicel 6–7 mm long; style 1–1.2 mm long; ovary ± 1 mm long.

KENYA. Machakos District: Kiboko area, 21 Apr. 2000, *P.A. & W.R.Q. Luke* 6199!; Masai District: Emali, 7 Mar. 1940, *van Someren* 27! & Chyulu Hills, 2°40'S 37°52'E, 19 Oct. 1969, *Gillett & Kariuki* 18859!
TANZANIA. Moshi District: on steppes near Lake Chala, *Volkens* 1806!
DISTR. **K** 4, 6; **T** 2; not known elsewhere
HAB. Grassland, bushed grassland and forest margins; 950–1900 m

SYN. *Asparagus nudicaulis* Baker in F.T.A. 7: 428 (1898)

NOTE. The other subspecies, subsp. *denudatus*, differs in having the branchlets almost perpendicular to the main stem; shorter internodes, cladodes, pedicel and style long; and a larger ovary. It occurs in South Africa.

11. **Asparagus flagellaris** (*Kunth*) *Baker* in J.L.S. 14: 614 (1875); F.W.T.A. 3: 93 (1968); U.K.W.F.: 311 (1994); Sebsebe Demissew in Fl. Somalia 4: 26 (1996) & Fl. Eth. & Erit. 6: 70 (1997). Type: Senegal, near Richard Toll, *Lelievre* s.n. (P!, holo.)

Erect shrub to 2(–3) m high, rarely climbing to 4 m high; branches terete or grooved, smooth to lined with straight or curved spines 3–5 mm long, these also on terminal branches, glabrous to pubescent. Cladodes in fascicles of 4–10(–14), subulate, stiff, 5–30(–60) mm long. Flowers axillary, solitary or paired; bracts overlapping, ovate, 1.5 × 1 mm, acute at apex; pedicels 5–10 mm long, articulated below the middle, closer to the base. Tepals white to purple (pink), ± equal, 2–3 mm long; stamens shorter than the perianth; anthers white or cream, 0.3 mm long; ovary 3-locular with 1–2 ovules in each locule; style ± 1 mm long, slender; stigma 3-branched. Berry orange-red, 6–9 mm in diameter with 1(–3) seeds. Fig. 2: 1–5 (page 16).

UGANDA. Karamoja District: Moruangaberu, 16 Feb. 1957, *Dyson-Hudson* 160!; Bunyoro District: 10 km Hoima–Masindi road, 9 Feb. 1971, *Kabuye* 315!; Masaka District: Mbara–Masaka road, 12 Aug. 1960, *Paulo* 682!
KENYA. Northern Province: Dandu, 22 Mar. 1952, *Gillett* 12615!; Machakos District: km 324 on main Nairobi–Mombasa Road, 29 Aug. 1959, *Verdcourt* 2368!; Tana River District: Kora Nature Reserve, 2 Aug. 1983, *Waterman* 1093!
TANZANIA. Hanang District: near Bulu village, 13 Feb. 1974, *Arasululu* 28842!; Singida District: near Matalele [Matelale], 11 Aug. 1927, *Burtt* 749!; Kilwa District: Selous Game Reserve, Kingupira, 25 Dec. 1977, *Vollesen* in MRC 4830!
DISTR. **U** 1–4; **K** 1, 3–7; **T** 1–8; West Africa to Ethiopia and Eritrea and S to Congo and Tanzania
HAB. Grassland, wooded grassland, scattered tree grassland, woodland, bushland, often in regularly burned areas; 0–2100 m

SYN. *Asparagopsis flagellaris* Kunth, Enum. Pl. 5: 103 (1850)
  Asparagus abyssinicus A.Rich., Tent. Fl. Abyss. 2: 319 (1851). Type: Ethiopia, Tigray, Djeladjeranne, *Schimper* III: 1479 (P!, holo.; BM!, iso.)
  A. pauli-guilelmi Solms in Schweinf., Beitr. Fl. Aethiop. 203 (1867). Type: Sudan, Akaro in Fesoghlu, *Herzog & Wurtemberg* s.n.; Kassan in Fesoghlu, *Cienkowsky* s.n. (both syn. not seen)
  A. schweinfurthii Baker in J.L.S. 14: 616 (1875). Type: Sudan/Ethiopia border, Galabat on the banks of the river Gendua, *Schweinfurth* 29 (K!, holo.)
  A. africanus Lam. var. *abyssinicus* (A.Rich.) Fiori in Boschi Piante Legn. Eritrea, 106 (1910)
  A. somalensis Chiov., Result. Sc. Miss. Stef.-Paoli, Coll. Bot. 1: 170 (1916). Type: Somalia, Aden Caboba, *Paoli* 924 (FT!, holo.)

NOTE. Some specimens are almost without cladodes during anthesis.
  A. flagellaris was previously distinguished from *A. schweinfurthii* Baker on the basis of the length of the cladodes, *A. flagellaris* having shorter cladodes, 1–2 cm long and *A. schweinfurthii* Baker having longer cladodes, 2.5–6 cm long or so. However, there are specimens from West Africa that previously were identified as *A. schweinfurthii* having cladodes that range from 1.5–6 cm long. Thus, in the absence of additional morphological characters to distinguish the two taxa, they are considered here to be conspecific.
  The specimen *Vollesen* MRC 4830 (K) from **T** 8 looks slightly different from the other specimens of the species, resembling *A. vaginellatus* Baker from Madagascar. On closer examination, it appears that *A. vaginellatus* in comparison to *A. flagellaris* has solitary, rarely paired flowers and fruits ± 4–5 mm in diameter. In contrast, *A. flagellaris* has paired flowers per fascicle (rarely solitary) and fruits 6–9 mm in diameter. Thus the specimen *Vollesen* MRC 4830 is treated as *A. flagellaris*. However, it appears that *A. vaginellatus* is closely related to *A. flagellaris* (particularly with specimens previously treated as *A. pauli-gulielmi* with scandent branches). Whether the two species are conspecific or not requires more field observation and study.

12. **Asparagus schroederi** *Engl.* in E.J. 32: 97 (1903); F.W.T.A. 3: 93 (1968); Obermayer in F.S.A. 5(3): 53 (1992). Type: Togo, Sokode, *Schröder* 20 (B†, holo.)

Erect shrubs 0.5–1 m high, rigid, spiny, from a woody rootstock; stems many, striate, the ridges bearing minute, sharp, transparent cells, rarely smooth, brown when young, grey with age branches simple, bearing overlapping cladodes in fascicles; spines on main stems sharp, 4–10 mm long, straight to recurved; spinules present below each fascicle, 3–4.5 mm long. Cladodes 3–10 per fascicle, filiform (somewhat triangular in cross section), 20–40(–65) mm long, turning dark when dry, deciduous, absent at time of flowering. Inflorescence: flowering branchlet 2.5–15 cm long or more; flowers usually paired, placed close together along branchlets; bracts membranous, fimbriate, ovate, ± 2.5 mm long; pedicels ± 2 mm long, terminating in a swollen disc below the perianth. Tepals white or cream, 3–4.5 mm long; ovary ovoid, 3-locular with 4 ovules in each locule; style ± 1 mm long with 3-branched stigma. Berry red, 6–8 mm in diameter, 1–2-seeded.

TANZANIA. Handeni District: Kideleko, 15 July 1982, *Leliyo* 220!; Ufipa District: Chapota, 22 Nov. 1949, *Silungwe* 28!; Iringa District: Kikobwe, 30 June 1965, *Mwakayoka* 13235!
DISTR. **T** 1–8; Togo, Cameroon, Congo (Kinshasa), Burundi, Angola, Zambia, Malawi, Zimbabwe, Namibia, Botswana and South Africa
HAB. Deciduous woodland, grassland, bushland, forest edge; 400–1850 m

SYN. *A. striatus* De Wild. in F.R. 12: 293 (1913), *non A. striatus* Thunb. Type: Congo (Kinshasa), Upper Shaba, Kakonde, *Hock* s.n. (BRU!, holo.)
   A. wildemanii Weim. in Bot. Notis. 1937: 446 (1937), *nom. nov.* for *A. striatus* De Wild.
   A. aspergillus sensu Jessop in Bothalia 9: 71 (1966) for specimens cited from Namibia
   Protasparagus schroederi (Engl.) Oberm. in S. Afr. Journ. Bot. 2: 244 (1983)

NOTE. The specimen *Gobbo & Zacharia* 141 from Kitwe Forest near Lake Tanganyika has much longer cladodes (50–65 mm long) than usual, but otherwise falls within the variation in the species.

13. **Asparagus racemosus** *Willd.*, Sp. Pl. 2: 152 (1799); Wight, Ic. Pl. Ind. Or.: t. 2056 (1853); Baker in J.L.S. 14: 623 (1875); Engler, Hochgebirgsfl. trop. Afr.: 169 (1892); Baker, F.T.A. 7: 434 (1898); Jessop in Bothalia 9: 72 (1966); F.W.T.A. 3: 93 (1968); Blundell, Wild Fl. E.Afr.: fig. 246 (1987); U.K.W.F.: 311, t. 140 (1994); Sebsebe Demissew in Fl. Somalia 4: 26 (1995) & Fl. Eth. & Erit. 6: 71 (1997). Type: India orientalis, in herb. Willdenow (B-W!, holo.)

Climbing shrub to 7 m high; branches terete, lined or angled, glabrous with spines 2–3 mm long in young parts to 5–8 mm long in older parts. Cladodes in fascicles of 2–6, subulate to flattened, 8–35(–40) × 0.5–0.7 mm. Inflorescence racemose, 1.5–19 cm long, glabrous; racemes solitary or fascicled, branching; the terminal parts of racemes with umbellate fascicles of 4–6 flowers; bracts ovate, concave, 1.5–4 mm long, glabrous, membranous, sometimes falling quickly; pedicel 4–8 mm long (elongating to 10 mm long in fruit), articulated at the middle or below. Tepals greenish white to white, 2.5–3 mm long; stamens shorter than the perianth parts with anthers orange to red; ovary obovate, 3-locular with 6–7 ovules in each locule; style 1–1.25 mm long with 3-branched stigma. Berry green turning red at maturity, 8–10(–13) mm in diameter, commonly 1-seeded, sometimes 2–3-seeded.

UGANDA. Acholi District: Imatong Mountains, Agoro, no date cited, *Eggeling* 1166!; Toro District: Ruwenzori, June 1893–94, *Scott-Elliot* 7846!; Mbale District: Elgon, Busano, 21 Dec. 1926, *Snowden* 1036!
KENYA. Northern Frontier District: Mt Kulal, 8 Oct. 1947, *Bally* 5516!; Londiani District: Tinderet Forest Reserve, 12 July 1949, *Maas Geesteranus* 5474!; Narok District: Nasampolai valley, 12 Oct. 1969, *Greenway & Kanuri* 13841!
TANZANIA. Masai District: Ngorongoro Crater, Empakai, 22 June 1973, *Frame* 159!; Handeni District: Sindeni village, ± 10 km W of Sindeni, 26 Nov. 1979, *Hedberg et al.* TMP 240!; Ufipa District: Kawa R. gorge, 30 Dec. 1956, *Richards* 7403!
DISTR. **U** 1–4; **K** 1, 3–6; **T** 1–4; Sudan, Ethiopia, Eritrea, Somalia, Zimbabwe; Asia
HAB. Forests, forest margins, secondary forest, wooded grassland, secondary bushland and grassland; (350–)1550–2900 m

SYN. *Asparagopsis floribunda* Kunth, Enum. Pl. 5: 98 (1850), *nom. illegit.* Type as for *A. racemosus*.
   Asparagus petitianus A.Rich., Tent. Fl. Abyss. 2: 320 (1851). Type: Ethiopia, Tigray, Mt Semajata, *Schimper* I: 374 (K!, iso.)
   A. racemosus Willd. var. *longicladodius* Chiov. in Malpighia 34: 530 (1937). Type: Ethiopia, Gojjam, Moyat in Agaumedir, *Taschdjian* 273 (FT!, holo.)
   Protasparagus racemosus (Willd.) Oberm. in S. Afr. Journ. Bot. 2: 244 (1983) & in F.S.A. 5(3): 45 (1992)

NOTE. The species resembles *A. buchananii*, but is easily distinguished by the following character combinations: *A. racemosus* has pedicels articulated at the middle or below; style 1–1.25 mm long; inflorescence commonly branched and spines on main branches not more than 8 mm long. In contrast, *A. buchananii* has pedicels articulated at the apex (base of perianth), inflorescence unbranched and spines on main branches 10–30(–40) mm long.

14. **Asparagus leptocladodius** *Chiov.* in Atti Reale Acad. Ital. Mem. Cl. Sci. Fis. 9: 58 (1940); Sebsebe Demissew, Fl. Somalia 4: 26 & Fl. Eth. & Erit. 6: 71 (1997). Type: Ethiopia, Bale, El Marra, Mt Ellot, 6°42'N, *Reghini* 8 (FT!, holo.)

Erect or scandent shrub to 2 m high; branches terete, white, peeling, glabrous to puberulous, with erect spines 4–12 mm long below cladodes. Cladodes in fascicles of 2–15, arcuate, 1–6 cm long, triangular, caducous during anthesis. Inflorescence racemose; raceme 0.5–2 cm long, often condensed and reduced, giving impression of an umbel; bracts ovate, 1–1.5 × 0.5 mm, white, falling quickly; pedicel 5–6 mm long, articulated in the middle or below. Tepals white to cream, ± equal, 3–4 mm long; stamens shorter than the perianth; anthers black; ovary 2–3-locular with 6–8 ovules in each locule; style 0.3–0.7 mm long with 2–3-branched stigma. Berry red, 5–9 mm in diameter, 1-seeded.

KENYA. Northern Frontier: S end of Huri Hills, 25 Feb. 1963, *Bally* 12350!; Tana River District: Tana River National Primate Reserve, main gate, 20 Mar. 1990, *Luke et al.* TPR 741!; Kilifi District: Watamu, 26 Apr. 1974, *Ngweno* 7C!
DISTR. **K** 1, 7; Ethiopia, Djibouti and Somalia
HAB. *Acacia–Commiphora* scrub, scattered tree grassland, bushed grassland; 0–750 m

SYN. *A. racemosus* Willd. var. *ruspolii* Engl. in Ann. Ist. Bot. Roma 9: 245 (1902). Type: Ethiopia, Bale, between Elb & Web Rivers, *Ruspoli & Riva* 771 [851] (521) (FT!, syn.)
   A. gillettii Chiov. in K.B. 1941: 183 (1941). Type: Somalia, 'Elmis', *Gillett* 4501 (K!, holo., FT!, iso.)

15. **Asparagus aspergillus** *Jessop* in Bothalia 9: 71 (1966); Sebsebe Demissew in Fl. Somalia 4: 26 (1995) & in Fl. Eth. & Erit. 6: 71 (1997). Type: South Africa, Letaba District, Birthday Gold Mine, *Breyer TM* 19063 (PRE!, holo.)

Climbing or erect herb or shrub to 2 m; branches glabrous to scabrid, pale grey, with spines 3–10 mm long. Cladodes in fascicles, subulate, 7–20 mm long and < 0.5 mm in thickness, absent at the time of flowering. Inflorescence racemose; raceme solitary or paired, 1.2–4.5 cm long, scabrid; bracts ovate, 0.5–1 mm long; pedicels solitary or paired, 1.5–2.5 mm long, articulated at the apex. Tepals white to cream, oblong to obovate, ± 3 mm long, ± equal; stamens 6, slightly shorter than the perianth parts; anthers black; ovary 3-locular with 4–6 ovules in each locule; style 0.75–1 mm long, 3-branched. Berry red, globose, ± 6 mm in diameter, 1–2-seeded.

KENYA. Northern Frontier District: Moyale, 14 July 1952, *Gillett* 13590!; Machakos District: Makipenzi, 9 June 1954, *Bally* 9731!
TANZANIA. Mbulu District: 2 km NE of Magugu, 14 July 1964, *Welch* 606!; Pare/Lushoto District: Mkomazi Game Reserve, 4 June 1996, *Vollesen* 96/30!; Handeni District: Kwamkono, 10 Nov. 1990, *Archbold* 3334!
DISTR. **K** 1, 2, 4, 4/6, 7; **T** 2, 3; Ethiopia, Somalia southwards to South Africa
HAB. *Acacia–Commiphora* bushland, wooded or bushed grassland, dry grassland; 600–1600 m

SYN. *Asparagus racemosus* sensu Sölch, Beitr. Fl. Südwest-Afr. 38 (1961)
   Protasparagus aspergillus (Jessop) Oberm. in S. Afr. Journ. Bot. 2: 243 (1983) & in F.S.A. 5(3): 56 (1992)

NOTE. The species resembles *A. rogersii* but is easily distinguished by the subulate cladodes 7–20 mm long. *A. rogersii* has shorter cladodes, 4–5 mm long and a thicker inflorescence. *Carter & Stannard* 93 from **K** 2 is placed under this species. But pedicels seen have two articulations at the basal and upper parts, and flowers persistent, both characters unusual for the species. More material from the locality is needed.

16. **Asparagus rogersii** *R.E. Fries* in Wiss. Ergeben. Schwed. Rhod.–Kongo-Exped. 1: 230 (1916). Type: Zambia, Chirukutu by Kabwe [Brocken Hill], *Rogers* 8346 (K!, holo)

Stiff erect plant 30–75 cm high, not branching; old stems grey to whitish, striated, puberulous; young parts greenish, angled, glabrous to glabrescent; spines on main stems below branchlets, 1–10 mm long, spinules below cladodes and flowers absent. Cladodes solitary or in fascicles of 2–13, flattened, 5–15 × 0.5–1 mm. Raceme simple, solitary or in fascicles of 2, 1.5–7 cm long; inflorescence axis thickened, slightly winged, ± 0.5 mm wide, flowers solitary or in pairs; bracts ovate, 0.5–1 mm long, ± leafless during flowering; pedicels 0.5–1 mm long, articulated at base of perianth, with a disc. Tepals cream, broadly elliptic, 2–2.5 × 1.5 mm; ovary ovoid with tuberculate surface, 3-locular with 4–5 ovules in each locule; style ± 0.5 (incl. stigma), 3-lobed. Fruit not seen.

KENYA. West Suk District: near Kongelai, July 1964, *Tweedie* 2859!
TANZANIA. Morogoro District: about 26 km E of Morogoro, 25 Nov. 1955, *Milne-Redhead & Taylor* 7375!
HAB. Dry wooded grassland; secondary *Brachystegia* woodland; 450–1500 m
DISTR. **K** 2; **T** 6; also in Zambia

17. **Asparagus buchananii** *Baker* in K.B. 1893: 211 (1893) & F.T.A. 7: 434 (1898); Jessop in Bothalia 9: 67 (1966); Sebsebe Demissew in Fl. Somalia 4: 27 (1995) & Fl. Eth. & Erit. 6: 71 (1997). Type: Malawi, Shire highlands, *Buchanan* 1503 (K!, lecto.)

Climber or scandent shrub to 3 m high; branches pale brown, smooth, glabrous, shiny with spines 1–4 cm long, dorsally flattened towards the base. Cladodes in fascicles of 3–5, subulate, 10–17(–27) mm long. Racemes simple, solitary or in fascicles of 2, 1.5–3(–5) cm long, glabrous; bracts ovate, 1–2 mm long; pedicels solitary, 2–3 mm long, articulated at the apex or sometimes at the middle. Tepals white to cream, 2–3 mm long; stamens shorter than the perianth parts; anthers yellow; ovary 3-locular, obovate with 6–8 ovules in each locule; style ± 0.5 mm long with 3-branched stigma. Berry red, 5–7 mm in diameter, 1–2-seeded.

UGANDA. Karamoja District: Lodoketemit [Lodoketeminit], 21 Feb. 1963, *Kerfoot* 4817!; Ankole District: 3 km E of Kamatarisi, 25 Sep. 1969, *Lye, Faden & Evans* 4322A!; Mbale District: Elgon, Kyosoweri [Kyasoweri], 13 Apr. 1929, *Snowden* 1064!
KENYA. Northern Frontier District: Lorogi [Lerogi] Forest, 28 km N of Maralal, 15 Nov. 1977, *Carter & Stannard* 411!; Masai District: 9 km N of Oloitokitok, 10 Mar. 1977, *Hooper & Townsend* 1300!; Kilifi District: 3 km E of Ganze, 15 Sep. 1985, *Robertson* 4040!
TANZANIA. Lushoto District: W Usambara, Mombo–Soni road, 24 June 1953, *Drummond & Hemsley* 3009!; Tabora District: Simbo Forest Reserve, 13 Oct. 1971, *Ruffo* 431!; Masai District: 8 km E of Masasi, Mkwera Hill, 16 Mar. 1991, *Bidgood, Abdallah & Vollesen* 2015!
DISTR. **U** 1–3; **K** 1, 3–7; **T** 1–8; Congo (Kinshasa), Rwanda, Burundi, Sudan, Ethiopia, Angola, Zambia, Malawi, Zimbabwe and South Africa
HAB. Grassland, wooded grassland, dry bushland, thicket, woodland; (0–)150–2250 m

SYN. *Protasparagus buchananii* (Baker) Oberm. in S. Afr. Journ. Bot. 2: 243 (1983) & F.S.A. 5(3): 45 (1992)

NOTE. The species is related to and resembles *A. racemosus*, but is easily distinguished as shown under the note in the latter species.
  Specimens from Tanzania, Usambara (*Holst* 3488) and Rungwe District (*Richards* 9878) show branching inflorescences (as in *A. racemosus*), but with the pedicel articulated above the middle, purplish stems and larger thorny spines as seen in other specimens of *A. buchananii*. However, these specimens might represent hybrids between *A. racemosus* and *A. buchananii*.

18. **Asparagus falcatus** *L.*, Sp. Pl. 1: 313 (1753); Bresler, Dissert. inaug.: 2 (1826); Kunth, Enum. Pl. 5: 71 (1850); Baker, Journ. Linn. Soc. 14: 626 (1875) & Fl. Cap. 271 (1896); U.O.P.Z.: 133 (1949); F.P.U.: 203 (1962); Jessop in Bothalia 9: 69 (1966); Blundell, Wild Fl. E.Afr.: fig. 534 (1987); U.K.W.F.: 311 (1994) pro parte; Sebsebe Demissew in Fl. Somalia 4: 26 (1995) & Fl. Eth. & Erit. 6: 70 (1996) pro parte. Type: *Burmann* s.n., Flora Zeylanica t. 13, f. 2 (1737).

Climbing or scandent shrub to 4(–7) m long. Branches with smooth, terete to angled stem, glabrous; spines recurved, on the main stem 5–7(–10) mm long, widened at the base; spinules at the base of cladodes 1–2 mm long. Cladodes in fascicles of 3–6, flattened, straight or falcate, (25–)35–125 mm × 2–5 mm, with a distinct mid-vein. Racemes simple, solitary or in fascicles of 2, 1.5–3(–5) cm long, glabrous; bracts ovate, 0.5–2 mm long; pedicels 1.5–4 mm long, articulated in the middle, above or below; flowers solitary. Tepals white to cream-yellow, broadly elliptic or obovate, 2.5–3.5 mm long; stamens shorter than the perianth; anthers yellow; ovary 3-locular with 6 ovules in each locule; style short, ± 0.5 mm long including stigma. Berry red or white flushed purple, 7–10 mm in diameter, 1–3-seeded.

UGANDA. Ankole District, 6 Sept. 1905, *Dawe* 470!; Mengo District: Bulwada [Balwada], Gomba, March 1932, *Eggeling* 596!; Masaka District: 1–2 km E of Kikoma, 19 Oct. 1969, *Lye* 4442!
KENYA. Navaisha District: 8 km N of Kinango, 25 Nov. 1958, *Moomaw* 1046! Kwale District: N of Jadini, 3 Dec. 1959, *Greenway* 9623!; Lamu District: Boni Forest, no date, *Adamson* 343 in *Bally* 5837!
TANZANIA. Tanga District: Amboni to Kibuguni, 25 Nov. 1936, *Greenway* 4781!; Uzaramo District: Dar es Salaam, Ubungo, 27 Nov. 1968, *Mwasumbi* 10416!; Lindi District: Rondo Plateau, 16 June 1995, *Clarke* 40!; Zanzibar, Pwani Mchangani, 26 Jan. 1929, *Greenway* 1201!
DISTR. **U** 2, 4; **K** 3, 7; **T** 1, 3, 6, 8; **Z**, **P**; Ethiopia, Somalia south to South Africa; also in Asia
HAB. Deciduous or evergreen bushland, thicket, dry forest and its margins; 0–850 m close to the coast and 1250–1350 m inland

SYN. *Asparagus falcatus* L. var. *falcatus* Bresler, Dissert. inaug.: 2 (1826); Kunth, Enum. Pl. 5: 71 (1850); Baker in J.L.S. 14: 626 (1875) & Fl. Cap. 6: 271 (1896); Jessop in Bothalia 9: 70 (1966)
  A. aethiopicus L. var *ternifolius* Baker in Saunders Ref. Bot.: t. 261 (1871). Type: South Africa, Natal, *Cooper* in *Saunders* 1448 (K!, holo.)
  Protoasparagus falcatus (L.) Oberm. in S. Afr. Journ. Bot. 2: 244 (1983)

NOTE. In a number of previous accounts (Jessop 1996; Sebsebe 1995, 1996), *A. falcatus* has been used in the wider sense. In this account, *A. falcatus* has been used in the strict sense to include taxa that were previously considered only to belong to *A. falcatus* var. *falcatus*. Most specimens of the species from East Africa (except *Festo* 777) have their pedicels articulated above the middle almost below the perianth, while specimens of *A. falcatus* from other areas have the their pedicels articulated below the middle.

19. **Asparagus natalensis** (*Baker*) *Fellingham & N.L.Mey.* in Bothalia 25 (2): 208 (1995). Type: South Africa, Natal, Inanda, *Medley Wood* 1351 (K!, holo.)

Climbing or scandent shrub to 2.5–3 m high; branches angled when young, becoming terete with age, glabrous to puberulous; spines 5–7 mm on main branches; spinules below the flowers 1–3 mm long. Cladodes solitary or in fascicles of 2–6, flattened, 15–25 × 1–3 mm, acute at the apex, attenuate at the base. Inflorescences (modified flowering branchlets often with cladodes or compound racemes), 1.5–15 cm long. Flowers in fascicles of 2–6; bracts ovate, 1.5 × 1 mm, acute at apex; pedicel 3–4 mm long, articulated in the middle or below. Tepals white to cream, ± 3 mm long, outer segments ciliate at the margin; stamens included in the perianth, anthers yellow; ovary 3-locular with 4–6 ovules in each locule. Berry red, globose, 9–10 mm in diameter, 1-seeded.

UGANDA. Ankole District: Karagwe, near Kagera River, Aug. 1893/94, *Scott Elliot* 8146!
KENYA. Northern Frontier District: Mt Kulal, 8 Oct. 1947, *Bally* 5517a!; Kiambu District: Kikuyu escarpment, Lari Forest Reserve, 22 Oct. 1972, *Hansen* 738!; Masai District: Eorengitok [Orengitok], 19 km from Narok, 17 May 1961, *Glover, Gwynne & Samuel* 1254!
TANZANIA. Arusha District: Arusha National Park, Small Momela Lake, 24 Aug. 1970, *Richards* 25778!; Masai District: Ngorongoro, 9 Oct. 1977, *Raynal* 19506!; Pare District: Upare, Oct. 1927, *Haarer* 879!
DISTR. **U** 2; **K** 1, 3, 4, 6; **T** 2, 3; Ethiopia and Somalia, South Africa
HAB. Dry bushland, dry forest and forest margins, wooded grassland, bushed grassland, thicket; (900–)1500–2300(–2700) m

FIG. 2. *ASPARAGUS FLAGELLARIS* — **1**, branch with flowers, × $^2/_3$; **2**, branch with cladodes and fruits, × $^2/_3$; **3**, cladode, × 4; **4**, pedicel with flower, × 2; **5**, fruit, × 2; **6**, seed, × 2. *ASPARAGUS ARIDICOLA* — **7**, branch with flowers, × $^2/_3$; **8**, fruiting branch, × $^2/_3$; **9**, detail of stem, × 4; **10**, cladodes, × 1; **11**, fruit, × 2; **12**, seed, × 2. 1, 4 from *Richards* 19201; 2, 3, 5, 6 from *Siame* 44. 7, 9, 10 from *Drummond & Hemsley* 4272; 8, 11, 12 from *Gilbert & Sebsebe* 8789. Drawn by Juliet Williamson.

SYN. *Asparagus aethiopicus* L. var. *natalensis* Baker in Fl. Cap. 6: 272 (1896)
    A. falcatus L. var. *ternifolius* sensu Sebsebe (1995), *non* sensu Jessop (1966)
    Protasparagus natalensis (Baker) Oberm. in S. Afr. Journ. Bot. 2: 244 (1983)

20. **Asparagus aridicola** *Sebsebe* **sp. nov.** *A. natalensi* Baker similis sed inflorescentiis simplice racemosis (non e ramulis modificatis ortis), pedicellis supra (non infra) medio articulates, segmentis periantii ciliates differt. Type: Ethiopia, Sidamo, 12 km NW of Moyale on road to Mega, *Gilbert & Sebsebe Demissew* 8789 (ETH!, holo.; K!, iso.)

Climbing or scandent shrub 1–3 m high; branches terete, sometimes peeling to expose brownish stems, glabrous to puberulous; spines 3–8 mm long on main branches, below branchlets and or cladode fascicles, 1–3 mm long on terminal branches below the flowers and cladodes. Cladodes solitary or in fascicles of 2–6, flattened, 10–30 × 1–2 mm, acute at the apex, attenuate at the base. Inflorescences simple racemes, 1.5–3.5 cm long; flowers solitary or paired; bracts ± 0.5 mm long; pedicel 3–4 mm long, articulated above the middle. Tepals white to cream, 3–4 mm long, smooth at the margin; stamens included in the perianth, anthers yellow; ovary 3-locular; 4–5 ovules in each locule. Berry red, globose, 6–7 mm in diameter, 1–2-seeded. Fig. 2: 7–11 (page 16).

KENYA. Northern Frontier District: Dandu, 12 Apr. 1952, *Gillett* 12770; Laikipia District: Aberdare Forest beyond Ndaragwe, 14 Mar. 1977, *Hooper & Townsend* 1345!; Teita District: Maungu–Ndara Road, 12 Sept. 1953, *Drummond & Hemsley* 4272!
DISTR. **K** 1, 3, 4, 7; Ethiopia
HAB. *Acacia–Commiphora* bushland, *Acacia* desert scrub, wooded grassland and thicket; 400–900 m

SYN. *A. africanus* var. *pubescens* Chiov. in Webbia 8: 9 (1951). Type: Ethiopia, Asile, near Meno River, *Corradi* 4653 (FT!, lecto.)
    A. falcatus L. var. *ternifolius* sensu Sebsebe (1996) pro parte
    A. aridicolus Sebsebe in Sebsebe, Nordal & Stabbetorp, Flowers of Ethiopia: Aloes and other Lilies: 192 (2003), *nom. nud.*

NOTE. *A. aridicola* is related to *A. natalensis*, but can be distinguished by the simple raceme inflorescence, pedicels articulated above the middle, and outer perianth segments entire at the margin. In contrast, in *A. natalensis*, the inflorescence is a modified branchlet, pedicels are articulated below the middle and outer perianth segments are ciliate at the margin.

21. **Asparagus faulkneri** *Sebsebe* **sp. nov.** *A. falcati* L. cladodiis complanatis similis sed cladodiis brevioribus, 4–7 (non 20–135) mm longis, pedicellis 1–2 (non 2–5) mm longis differt. Type: Tanzania, Zanzibar, Mazizini [Massazime], *Faulkner* 2375 (K!, holo.)

Climber; stem yellowish (on drying), furrowed longitudinally, glabrous; young branches striate; spines on main branches and branchlets straight, 3–8 mm long, broader at the base. Cladodes in fascicles of 4–8, linear to falcate, 6–15 × ± 1 mm, acute at the apex, narrowing towards the base. Inflorescence simple, 1.5–2.5 cm long; bracts ovate, 1 × 1 mm; pedicels 1–1.5 mm long, articulated just below the perianth. Tepals white to cream, 2–3 × 1 mm; stamens shorter than the perianth, ± 2 mm long including anthers; anthers orange; ovary 3-locular with 4 ovules in each locule; style ± 0.5 mm long including stigma, 3-branched. Berry red, 5–7 mm in diameter, 1-seeded. Fig. 3: 5–7 (page 18).

KENYA. Kwale District: Between Kinango & Mariakani, 23 Nov. 1971, *Bally & Smith* B14364!; Kilifi District: Mida Creek, Mida, 10 Jan. 1970, *Bjornstad* 211!
TANZANIA. Lushoto District: Segoma, 23 Mar. 1964, *Faulkner* 3924!; Pangani District: Mwera, Pangani, 24 Aug. 1957, *Tanner* 3644!; Lindi District: Lake Lutamba, 9 Oct. 1934, *Schlieben* 5453! Zanzibar, 1927, *Toms* 125!

FIG. 3. *ASPARAGUS USAMBARENSIS* — **1**, flowering branch, × ²/₃; **2**, branch with cladodes, × ²/₃; **3**, cladodes, × 2; **4**, pedicel with flower, × 6. *ASPARAGUS FAULKNERI* — **5**, flowering branch, × ²/₃; **6**, base of 2° order branch, × 4; **7**, pedicel with flower, × 6. 1–4 from *Semsei* 2706. 5–7 from *Faulkner* 2375. Drawn by Juliet Williamson.

HAB. Coastal bushland, dry forest and wooded grassland; 0–250(–450) m
DISTR. **K** 7; **T** 2–3, 8; **Z**; not known elsewhere

NOTE. The species is affiliated with *A. falcatus*, but has much shorter cladodes 5–7 × 1 mm (not 35–125 × 2–5 mm); flower pedicels are 1–2 (not up to 4) mm long.

22. **Asparagus usambarensis** *Sebsebe* **sp. nov.** *A. falcate* L. similis sed cladodiis 1–1.5 (non 2–5) mm latis, inflorescentiis 4.5–10 (non 1–3) cm longis, fructibus ± 6 (non 7–10 mm) diametro facile distinguenda. Type: Tanzania, Lushoto District, W Usambara, Shume Forest Reserve, *Semsei* 2706 (K!, holo.)

Climbing, scandent or procumbent perennial herbs, spiny; young stems striate, older ones terete, glabrous; spines on the main stem 5–8 mm long, sharp; spinules at the base of cladodes 0.5–1.5 mm long. Cladodes fasciculate, solitary or in fascicles of 2–3, flattened, straight or falcate, 10–30(–40) mm × 1–1.5 mm, acute at the apex, narrowed at the base. Inflorescence racemose, solitary or in fascicles of 2 together, 4.5–10 cm long, glabrous; bracts ovate or lanceolate, 1.5 × 0.5 mm, acute at apex; pedicels 2.5–6 mm long, articulated near the middle or slightly below. Tepals white to cream, 3–3.5 mm long; stamens shorter than the perianth; anthers orange; ovary 3-locular with 6 ovules in each locule; style ± 1 mm long including stigma. Berry red or white flushed purple, ± 6 mm in diameter, 1-seeded. Fig. 3: 1–4 (page 18).

TANZANIA. Lushoto District: Matondwe Hill at the head of Kwai valley, 28 Feb. 1953, *Drummond & Hemsley* 1338! & Mountain ridge between Matundsi and Mashindei Peaks SW of Anbangulu Tea Estate, 4 Feb. 1985, *Borhidi et al.* 85/464!; Tanga District: E Usambara, Mlinga Peak, 18 Feb. 1937, *Greenway* 4902!
DISTR. **T** 3; restricted to the Usambara Mts
HAB. Evergreen rain-forest, forest edge, thicket and in rock crevices on rocky slopes, (1000–)1200–1950 m

NOTE. The flowers are described as having an unpleasant, sweet rather sickly odour.
The species is related to *A. falcatus* but differs by the cladodes being 9–25(–40) × 1–1.5 mm; inflorescences 4.5–9 cm long and occurring between (1020–)1800 and 1950 m altitude. On the other hand, *A. falcatus* has cladodes 15–90 × 2–5 mm; inflorescences commonly 1–3 cm long and occurring between sea level and 1350 m altitude.
The species also resembles the south African *A. densiflorus* on account of its shorter cladodes and similar sized flowers. However, it is distinguished by its climbing to scandent habit, branches spread out (not condensed), 10–15 mm gap in between branches and having flexible cladodes. In contrast, *A. densiflorus* has an erect habit, branches aggregated with a gap of only 2–3 mm in between branches and having stiff cladodes.

subgenus MYRSIPHYLLUM

(*Willd.*) *Baker* in J.L.S. 14: 597 (1875)

Medeola L., Sp. Pl. 339 (1753)
Myrsiphyllum Willd. in Ges. Naturf. Fr. Berlin Mag. 2: 25 (1808); Kunth, Enum. Pl. 5: 105 (1850); Obermeyer in Bothalia 15(1 & 2): 77–78 (1984)
Asparagus section *Myrsiphyllum* (Willd.) Baker in Fl. Cap. 6: 258 (1896)
Hecatris Salisb., Gen. Pl. Fragm. 66 (1866)

23. **Asparagus asparagoides** (*L.*) *Wight* in Cent. Dict. II: 845 (1909); Salter in Fl. Cape Penins. 173 (1950); U.K.W.F.: 311 (1994); Jessop in Bothalia 9: 81 (1996); Sebsebe, Fl. Eth. & Erit. 6: 73 (1997). Type: Tilli, Catalogus Plantarum Horti Pisani, t. 12, f. 1 (1723) as 'Asparagus africanus scandens Myrti folio' (iconotype)

Climbing or suberect annual herb to 3 m high; branches terete or angled, glabrous, without spines. Cladodes broadly ovate to lanceolate, 1.2–8 × 0.7–3 cm, acute at the apex, rounded at the base, with numerous (>15) parallel lateral veins. Racemes solitary or 2 together; bracts ovate, membranous, ± 3 mm long; pedicel 5–32 mm long, articulated at the apex or in the upper half. Tepals greenish white, 5–6 mm long; stamens 6, ± 6 mm long, shorter than the perianth; ovary 3-locular with 4–6 ovules in each locule; style 1–3 mm long, stigma 3-branched. Berry red, globose, 6–10 mm in diameter, up to 8-seeded.

UGANDA. Kigezi District: E Virunga, 27 Oct. 1954, *Stauffer* 649!; Mbale District, Elgon, Bukalsi trail, 29 Mar. 1997, *Wesche* 1311!
KENYA. Northern Frontier District, Furrole Mt, 15. Sept. 1952, *Gillett* 13889!; Trans-Nzoia District: Kitale, Aug. 1971, *Tweedie* 4088!; Kiambu District: Ngong, 25 May 1931, *van Someren* 1156!
TANZANIA. Masai District: Lerong, W side of Lemagrut, 1 Feb. ?1960, *Newbould* 5639!; Kondoa District: Kinyassi escarpment, 7 Feb. 1928, *Burtt* 1727!; Rungwe District: N slope of Rungwe Mt, 8 Feb. 1961, *Richards* 14282!
DISTR. U 2, 3; K 1, 3, 4, 7; T 2–7; widespread in tropical Africa and extending to warmer parts in Europe. In recent years it has naturalised in Australia
HAB. Dry or moist forest and forest margins, riverine forest, less often in woodland or secondary bushland; 1100–2400(–3000) m

SYN. *Medeola asparagoides* L., Sp. Pl. 339 (1753)
    M. angustifolia Mill., Gard. Dict., 8th ed., n.2 (1768). Type: Tilli, Catalogus Plantarum Horti Pisani, t. 12, f. 2 (1723) as 'Asparagus africanus scandens Myrti folio angustiore' (icono.)
    Dracaena medeoloides L.f., Suppl.: 203 (1781). Type: South Africa, Cape of Good Hope without precise locality, *Thunberg* s.n. (UPS!, holo.)
    Myrsiphyllum asparagoides (L.) Willd. in Ges. Naturf. Fr. Berlin Mag. 2: 25 (1808); Kunth, Enum. Pl. 5. 105 (1850); Hook. f. in Bot. Mag.: t. 5584 (1866)
    Hecatris asparagoides (L.) Salisb., Gen. Plant.: 66 (1866)
    Asparagus medeoloides (L.f.) Thunb., Prodr.: 66 (1794); Marloth, Fl. S. Afr. 4: pl. 20 (1915).
    Myrsiphyllum angustifolium (Mill.) Willd. in Ges. Naturf. Fr. Berlin Mag. 2: 25 (1808)
    M. falciforme Kunth, Enum. Pl. 5: 107 (1850). Type: South Africa, Cape without precise locality, *Drége* 2704a in " Herb. Luc. " (K!, iso.)
    Asparagus medeoloides (L.f.) Thunb. var. *angustifolius* (Mill.) Baker in Fl. Cap. 6: 273 (1896).
    A. kuisibensis Dinter in F.R. 29: 270 (1931). Type: Namibia, Kuisib River, *Tjuezu* in Herb. Dinter 4698 (B!, holo.)

NOTE. The leaves from specimens in woodland and bushland tend to be smaller in size than from the forest e.g. *Bidgood et al.* 3435! and *Pocs & Chuwa* 890007/B!

UNCLEAR SPECIES

Asparagus buruensis Engl. in E.J. 45: 155 (1910). Type: Kenya, Teita District: between Taveta and Taita Hills [Buru Mts], *Engler* 1925 (B†, holo.)
Engler said this taxon was close to *A. irregularis*; this is now a synonym of *A. africanus*.
It is possible *A. buruensis* is a synonym of *A. africanus* but as the type has been destroyed this remains unresolved.

INDEX TO ASPARAGACEAE

New names validated in this volume

Asparagus africanus *Lam.* var. **puberulus** (*Baker*) *Sebsebe* **stat. & comb. nov.**
Asparagus aridicola *Sebsebe* **sp. nov.**
Asparagus denudatus (*Kunth*) *Baker* subsp. **nudicaulis** (*Baker*) *Sebsebe* **comb. & stat. nov.**
Asparagus faulkneri *Sebsebe* **sp. nov.**
Asparagus migeodii *Sebsebe* **sp. nov.**
Asparagus usambarensis *Sebsebe* **sp. nov.**

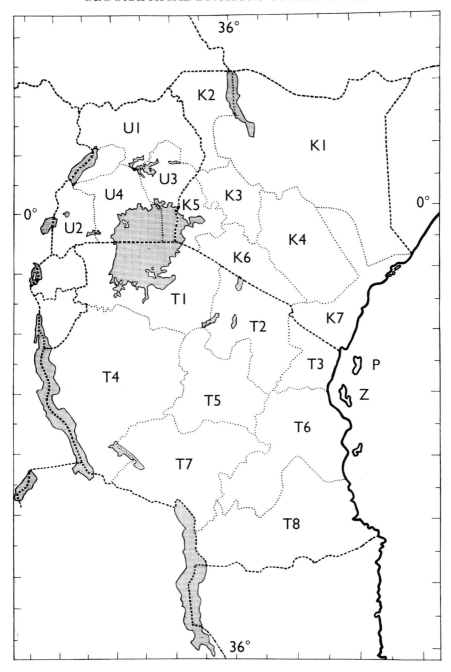

PLANTS PEOPLE
POSSIBILITIES

First published in 2006 by
Royal Botanic Gardens, Kew
Richmond, Surrey, TW9 3AB, UK
www.kew.org

ISBN 1 84246 116 8

Design by Media Resources, typesetting and page layout by Margaret Newman, Information Services Department, Royal Botanic Gardens, Kew.

Printed in the UK by Hobbs the Printers

For information or to purchase all Kew titles please visit www.kewbooks.com or email publishing@kew.org